DOUBLES DOMINATION
THE BEST OF THE BEST

TACTICS, TIPS, PRESSURE STRATEGIES,
SET PLAYS AND GREAT IDEAS FROM
THE BEST PROS IN THE WORLD

BOB ALLCORN

THE AUTHOR WITH COACHES FOR BRYAN BROTHERS,
DAVE MACPHERSON (RIGHT) AND DAVE MARSHALL (LEFT).

DEDICATION

This book is dedicated to all of the teaching pros in the world who have spent their lives training, practicing, learning and sacrificing to be the best they can possibly be and who now spend all of their time sharing their knowledge with wannabes, like me.

THANK YOU!!!

CONTENTS

INTRODUCTION

DEPLOYING WINNING DOUBLES TACTICS

Every day you have a choice: to work at your game and get better or do nothing and get worse. Your choice. Why not begin each day with the question: What can I do today to get better?

This book is not intended for the beginner, nor for the player above 4.5. It is intended for the serious adult doubles player who has more than a basic knowledge of the game and who loves to win at the club, USTA match and tournament level.

Except for a few basic reminders, it does not focus on strokes. It assumes that you have the strokes you have, and either will not be changing your strokes or will be changing them only through personal lessons with a pro. Rather, this book is about beating your opponents – gaining the upper psychological edge, using the USTA Rules to your advantage, knowing where to hit the ball and using sophisticated TACTICS to win. These TACTICS are as important as your strokes and should be practiced as you would practice your strokes.

This book is a synthesis and distillation of the very best of more than 1,000 hours of lessons (and more than $100,000 in fees) from the best pros around the world, the best ideas espoused on the internet and almost all of the books written about tennis – going all the way back to Bill Tilden in 1929. "Spalding's Tennis," William J. Tilden, 2nd (American Sports Publishing Co., NY, NY 1929)

It is written in the context of a right-handed player, assuming that lefties are intelligent enough to modify the instructions to their own use. It also assumes that gender identity is irrelevant, and no intent exists for any preference of "him" vs. "her" or vice versa.

WHAT YOU NEED TO WIN

MUST HAVES

There are a few shots and intangibles in doubles that are essential to know in order to utilize the tactics in this book. Practice them (refer to the "PRACTICE" section later in this book). These are:

- ✐ Willingness to take risks at the net – especially not being afraid of getting "beaten" down the line.

- ✐ A fairly accurate and consistent first serve – one that can be placed wide, down the middle or at the body.

- ✐ A reliable second serve, in which you have confidence. It doesn't need to be great or powerful or have lots of topspin – just reliable.
  This will prevent double-faults and will give you the confidence to be occasionally more aggressive with your first serve. When asked what the secret was to her world record 19 Master's doubles' titles (3 more than the Bryan Brothers) and more than two-year unbeaten winning streak playing doubles with Pam Shriver, Martina Navratilova responded that, "First of all, we both had good second serves." Probably the oldest adage about doubles is, "You are only as good as your second serve."

- ✐ A very reliable overhead. It doesn't need to be a blistering killer overhead that wins the point every time you hit it. Just reliable – even if just aimed behind the center "T".

- A GREAT lob. The most important shot in doubles and one of the most tactically underrated and underutilized. I am not suggesting that you become a lob king or queen – merely that you _own_ that shot and be able to use it – often.

- A good drop shot. Not necessarily great. Practice this so you can get the ball to bounce twice before it crosses your opponent's service line. See, "PRACTICE" section.

- A reliable volley that you can hit from _all_ points between the service line and the net. Most volleys, especially the serve and volley, occur closer to the service line than the net.

- Targets: Awareness of when to hit the ball to what target.

- Court sense – knowing where to be on the court when the ball, your opponents and your partner move.

VERY, VERY GOOD TO HAVE (BUT NOT ESSENTIAL)

- Slice forehand: a very offensive shot to nail your opponents to the baseline, to counter the serve-and-volley and to attack the mid-court players' feet.

- Topspin forehand: lots of net clearance; particularly effective when hit high and deep to your opponent's backhand.

- Slice backhand

- Serve and volley

- Half-volley: very important – particularly if you serve and volley.

- Soft volley (drop volley; stop volley)

- Slice serve: to take the deuce court returner wide off the court or to his body; also good for a down-the-middle serve to the AD court.

- Lob volley

- Deep serve: depth is better than speed or spin.

- Good attitude – external and internal, toward yourself and your partner

BONUS POINTS TO HAVE

- Topspin serve

- Topspin lob

- 90% accurate serve

- Topspin backhand

- Crushing overhead

- Underhand serve or other "junk" serve

- Swing volley

All of these will be discussed throughout this book.

GENERAL TIPS

THE GOAL

Your goal every point is to get to the net and stay at the net; and if you get pushed back, then get back to the net asap, keep your opponents away from the net and win the point at the net with a volley or overhead. By far the most points in doubles are won at the net. Not surprisingly, most points in doubles are lost from the back court.

There are six ways to get to net:

- Start there (or near there) as a net partner of the server or returner;

- Serve and volley;

- Approach shot;

- Sneak approach;

- Chip and charge;

- Run for a short shot by your opponents.

There are three ways to keep your opponents away from the net:

- Lob - either a chip lob or standard lob

- Keep the ball deep to them;

- Take over the net first.

As you can see, of those nine alternatives, you control eight of them. Learn all eight, practice all eight, become proficient at all eight and use all eight continuously and repeatedly in every match. And, while you are learning, learn how to offensively deal with short balls so that you control all nine options.

TEST AND EXPLOIT

Beginning with the warm-up (see WARM-UP, next section) and continuing throughout the match, continuously probe both opponents to discover their weaknesses. EVERYBODY HAS WEAKNESSES. Find them and then your first strategy becomes simple: attack those weaknesses mercilessly. Some examples – some obvious and some not:

- If one or both of your opponents are slow, make them run;

- If they like to stay at the baseline, hit drop-shots, short-angled shots and drop volleys;

- If they get too close to the net, lob them;

- If one has a weak backhand volley, pound shots at their backhand;

- If one doesn't move back well, lob her;

- If they love pace, hit them junk;

- If they have a great cross-court return, serve down the middle "T" to take away that shot or switch to the Australian formation (see Australian Formation section);

- If one has a weak second serve, move up to the service line to return it, intimidating them (see RETURNER section) and have your partner squeeze the center "T" (see RETURNER'S PARTNER section);

- If one has long arms or a two-handed backhand, then slice serves to their body.

THE "WARM-UP"

If possible, the traditional "warm-up" before matches should not be your actual warm-up. The intelligent player will have already warmed up their body, strokes and serve before the official "warm-up." The pros warm up for more than an hour before going on the court.

So, what should the "warm-up" be? A scouting session. Your main purpose during the warm-up is to study your opponents. What you want to look for:

- What are their strengths and weaknesses?

- What kinds of spin do they use?

- If you hit a ball right at them, do they take it with their forehand or backhand. Whatever they consistently choose is their favored, more reliable and more comfortable shot.

- How do they handle low slice shots?

- Do they bend their knees for low shots? For half-volleys?

- How do they handle deep topspin shots?

- Do they mostly slice their backhands? Do those float so as to be easily picked off at net?

- How do they move? Are they fast or slow?

- Do they move their feet in preparing for shots?

- How are their volleys? Consistent? Controlled?

- Which player is weaker at net? Which side is weaker?

- How close to the net do they like to stand?

- How are their overheads? Do they even practice their overheads during the warm-up? Not everybody does.

- What are their serves like? Pinpoint accurate or do they just land "wherever"?

- What spins do they use?

Learn as much as you can about the person you are warming up against, but also be looking at the opponent warming up with your partner so you get information about her game. And, before the match begins, discuss with your partner the detailed strengths and weaknesses of each opponent. Use that information to tweak your game plan. Maybe you want to hit balls right at the weaker opponent when he is at net. Maybe one of your opponents is slow and will have difficulty getting to drop shots? Maybe neither have practiced overheads so you should lob them to test them?

TACTIC: Insist on taking serves <u>during the warm-up</u> – not before each person serves their first service game. It is required by the USTA Code, Principle #4:

> *"A player should take all warm-up serves*
> *before the first serve of the match."*

Doing so also gives you the chance to return practice serves from both opponents while standing on your returning side of the court so you can get used to both opponents' serves (if you wait to practice serves until just before your opponent serves for the first time, then the only one to get a chance to return the practice serve will be the deuce court returner.)

TACTIC: Do you want to show your opponents your strengths? If you and your partner both have great volleys, should you show them off during the warm-up? Or does doing so merely encourage your opponents to engage in a lob fest against you to avoid your great volleys? A difficult decision, but you certainly do not want to show off your best shots during the warm-up, except your overhead.

TACTIC: It is a good idea to show off your overhead during warm-up, if you have a great one. If you can intimidate your opponents with your overhead even before the match begins, you will discourage them from lobbing you and give you the immediate chance to play closer to the net.

WATCH THE BALL? (NOT NECESSARILY/NOT ALWAYS)

CONSIDER THE FOLLOWING

- ~As the ball comes toward you from the other side of the net, do not watch "the ball." Rather, intensely focus on a small spot on the ball about the size of a quarter or, if you can discern any printing or color on the ball (or can see the seams), focus on that. You will need to practice this technique when not playing matches in order to make progress. But if you can succeed, you will find your rate of mis-hits dropping dramatically.

- ~Continue to watch the ball as it strikes your racquet, making sure you actually *see* the contact between ball and racquet. You should be able to see the yellow blur of the ball leaving your racquet (if not, then just notice the contact). Then, and only then, should you look up.

- ~ BUT, when you look up, do not follow the path of the ball across the net toward your opponent, as this deprives you of valuable information with which to be able to most quickly anticipate what kind of shot will be coming back at you. Rather, after seeing the yellow blur, immediately look at

your opponent's racquet. You will also peripherally be able to see the ball.

- Does your opponent take a big or a short backswing?

- Is the angle of their racquet open-faced, closed or neutral?

- Is the racquet below the net tape so that their shot will have to be hit up to get over the net?

- Are they stepping forward into the ball for an offensive shot or backward away from the ball for a defensive shot?

- Will they have to reach for the ball, resulting in a weaker shot?

Watching your opponent's racquet will give you the quickest and most accurate information as to whether the coming shot will be offensive, defensive or neutral; whether it will be a drive or a lob; whether it will be topspin or slice; or whether it will be a soft touch shot. This will give you the earliest indication about whether you need to move toward the net for a volley or move back to cover a lob.

BALL PLACEMENT

The single most important aspect of doubles is knowing where to hit the ball and with what kind of a shot. There are a multitude of shots to hit in most situations. Do *not* "just get the ball back" as some suggest. Your object is to take the net on every point (with a serve and volley, a chip and charge return or an approach shot) and force a weak shot by your opponents so that you can then hit a winner. So, in effect, every shot that is not intended to be a winner or a completely defensive shot should be an approach shot designed to get you and your partner to the net and to force a weak reply. Most doubles points do not last even 6 shots. If you can get the ball to go to where you know it should go and follow the ball to the net to put it away, you can and should be able to end most points with three of your own shots.

This book will identify (1) those locations on the court to aim for and (2) those shots to use to hit to those locations. Learn WHERE to hit the ball in every situation, HOW to get it there, and the NEXT STEP to take after getting it there. Practice this at least as much as you practice your strokes.

TIP: As mentioned in the PRACTICE section below, you can actually practice shot selection at home by watching a recorded doubles match, stopping the action just before a player hits the ball and thinking about what the best shot would be. Do not do this by watching mens' professional tennis as their game bears no resemblance to our game. Watch either womens' pro games or high-level non-professional matches. You can also access high-level advice on this by tuning into Brent Abel's YouTube broadcast "WTRS" ("What's the Right Shot"), where he does exactly that with senior mens' national championship points and even some complete matches. Analyzing what top players do and under what circumstances they do it is invaluable in helping to learn good shot selection. Up your shot selection savvy!

USE YOUR LEFT HAND

Watch the pros. They almost all use their left hand to hold the racquet, and many use their left index finger to rest on the strings at the same time. It is the very first thing that all pros at Peter Burwash International ("PBI") teach. This technique is *very valuable* and has many distinct and important benefits:

- Between points it enables you to rest your right hand and arm, which take a lot of strain and abuse during a match, and need rest;

- In preparing for a forehand volley, if you hold onto your racquet with your left hand until the forward movement of the stroke begins, it acts as a brake to keep the racquet from going behind you; and by keeping the racquet in front of you, it shortens your stroke and your reaction time to hit the ball.

- By turning to your side, the racquet will always be in the perfect position to volley: in front of you, the proper distance from your body, and with the head of the racquet up;

- Your left-hand acts as the perfect guide for holding the racquet exactly where it should be in preparation for a backhand volley or backhand dropshot.

- The latest methods for forehands and backhands include moving your left hand with your right one as you turn to prepare to hit the ball.

- As taught fervently by PBI pros (the world's largest high-end organization providing tennis pros to clubs and hotels around the world), you should not only hold the racquet with your left hand but also use your left index finger to place on the bottom of your racquet strings. This is a GREAT idea. It provides constant tactile feedback to your brain about the location of the racquet *and* the angle of its face. You can change grips instantaneously without using your right hand by merely "dialing in" the grip you want with your left hand – including quickly changing the grip to any weird one you may need at net to avoid getting hit in the face.

MOVE DIAGONALLY

When playing net, you must not move laterally (parallel to the net) as you move toward the ball. Almost all recreational players make the mistake of moving sideways. Instead, you must move diagonally - at an approximate angle of 45 degrees - usually toward the net strap, so that you are moving at an angle both laterally and toward the net at the same time. This movement:

- Is the shortest distance from you to the ball, enabling you to reach more balls;

- Cuts down the time it takes you to get to the ball and gets you closer to the net – both of which reduce the time your opponent has to react to your volley;

- Gets you closer to the net so that (1) when you volley you can get to the ball at its highest point (before it drops below the height of the net where you would have to hit the ball up to the opposing net player) and (2) you then have the best chance to hit down on the ball;

- Gets you closer to the net for more angles to be available for your volley; and,

- Provides forward momentum, which places more pace on the ball when struck.

Somewhat similarly, using a diagonal approach to get to the ball in the backcourt (especially when returning serve) cuts off the angle, gets you to the ball faster and adds to your stroke your body's forward momentum.

BEND THOSE KNEES (NOT YOUR BACK)

When asked what the single most important thing to do at home one could do to improve one's game, the world's most famous and most successful tennis coach, Nick Bollettieri, said to find a large mirror at home and get a piece of tape. In front of the mirror, get into a deep squatting position like an ice hockey goalie. Mark where the top of your head is on the mirror with the tape. Shadow your strokes in front of the mirror, making sure that your head <u>never</u> rises above the tape. Advice from the MASTER! Prepare low for your shots, make sure you are low on contact and <u>*stay low*</u> <u>*after the shot.*</u> For some unknown reason, the human tendency is to raise one's body (or at least one's head – which causes the body to rise) at the time of contact or shortly thereafter. This results in the racquet also being raised and not hitting the ball squarely. Take Nick's advice.

MOVE AFTER HITTING THE BALL

It is my habit (and that of everyone I know who is not a pro) to stand on the court with my racquet down, watching the shot I just made. Am I unsure whether it will go in? Am I admiring a "great" shot? Am I waiting to see what happens next? Am I waiting to see if my opponent gets to the ball? Am I thinking about dinner? Whatever the reason, it is clear what I am NOT doing: getting ready for the ball to come back. Immediately after hitting every shot you *must* prepare for the ball to come back. This means not only moving quickly into the best position and staying on your toes, but keeping your racquet up and in front of you. This applies to all shots, but particularly volleys, as the ball will come back much sooner than if one is at the baseline. So, my favorite pro forces me during practice to:

- ➋ After I chase down a wide shot, push off immediately with my outside leg to return to cover my assigned part of the court. I cannot expect my partner to cover the entire court. It needs to become a habit;

- After every volley, keep the head of my racquet up (gravity usually wins unless I am badgered); keep the racquet out in front of me; move quickly toward the ball so that (1) I intimidate the opposing net person to whom I have hit the ball; (2) I am as close to the net as possible so as to hit the ball at its highest point (and not below the level of the net) and give me the best array of angles possible; and (3) I take time away from my opponent by giving him less time to react to my shot.

I am trying to incorporate the automatic internal message of "HIT-MOVE" after every shot. Not easy to make automatic, but well worth every effort.

PLAYING AGAINST THE LOBBER

Opponents who constantly lob can drive one insane – but only if you don't have an effective plan that includes:

HIT SHOTS THAT ARE DIFFICULT TO LOB

The first defense against the lob is an offense! Hit balls that are the most difficult for your opponents to lob. If you keep hitting them easy shots to the middle of their court, you are feeding their habit. Make them run side-to-side and up-and-back, as well as hitting them low balls that are hard to get under. Use these tactical shots:

- Drop shots. Have you ever tried lobbing when you are running at full speed and have to reach for a low ball?

- Short angle shots to the side "T". Wide low shots make your opponents run <u>and</u> reach.

- Deep slices directly at their feet.

TAKE THEIR LOBS OUT OF THE AIR

One reason your opponents lob is to force you deep into your court behind the baseline. Moving in and volleying the ball before it bounces defeats that tactic, allows you to move toward the net instead of back away from it, and takes time away from your opponents to make their next shot. See, "TAKING THE BALL OUT OF THE AIR", next section.

IF THEY BOUNCE

If to your forehand side, you can either hit a serve or hit a high deep lob back at them. If to your backhand side, try to run around your backhand and hit it with your forehand. Either way: bend your knees and make sure your follow-through swing is high and your racquet face is open and pointing toward the sky when you complete your shot.

TAKE THE BALL OUT OF THE AIR

A huge part of winning doubles is putting pressure on your opponents and keeping it on them by getting to the net, staying on the offense and taking time away from them. One of the most effective ways to accomplish this goal is to take as many balls out of the air (from wherever you are in the court) as possible.

If you are at the baseline and your opponent lobs the ball high and deep so that it will bounce very high, do *not* let it bounce. Move in and hit it while it is still in the air. The alternative is to let it bounce, forcing you back toward the fence to hit a ball that is higher than your shoulders – placing you in as defensive a position as possible and allowing your opponents to come to net and be in an offensive position. You will lose the point. Instead of moving backwards, try moving *forward* to catch the ball in the air before it bounces and volley the ball back deep and high to your opponents.

By doing this, you will have hit a shot that (1) most opponents do not expect; (2) prevents your opponents from coming to net; (3) allows you to move forward and take over the net; (4) takes away a significant amount of time from your opponents and their preparation to hit the next shot; and (5) maintains the aggressive posture and pressure you want to display throughout the entire match. As an added benefit, you may have decimated their favorite strategy.

This tactic applies even to a deep lob hit over the head of your net partner, forcing you to cover the other side of the court. Try your best to cut off the ball at an angle and keep it from bouncing. Volley it as high and as deep as you can to stay in the point, to keep your opponents off the net and to give you and your partner time to recover your positions. If it is a difficult shot, aim cross-court as this is the deepest part of the court (4 1/2′ more court than down-the-line) and gives you the biggest target.

This tactic also works to counter a hard-hitting opponent who drives the ball deep to the baseline, forcing you to either move backward or hit half-volleys at your feet – neither of which is a great alternative. Try moving in to a spot between the service line and the baseline and take that shot as a low volley – also taking time away from your opponent. If the opponent is skillful and intelligent enough to counter this strategy, they will adjust their aim to still aim at your feet in your new position. You can respond to that strategy by standing in one spot and moving to the other (either up or back) as soon as the opponent begins to swing – as you would do for a poach.

EXCEPTION

The Bryan Brothers' (and other pros) sometimes handle the deep high bouncing lob another way: they let the ball bounce and then from the baseline they hit what is effectively a serve. This works if you have a great serve and can give the shot the added benefit of slice or even topspin. But remember: this shot can be picked off as well by an aggressive net opponent.

COVER THE MIDDLE

People are constantly taught: "Down the middle solves the riddle." A huge percentage of shots are hit "down the middle" of the court because –

- ➋ The net is lowest there;

- ➋ The target is the biggest and therefore the margin of error is the best;

- ➋ It "splits" your opponents, forcing them to decide which partner will take the ball;

- ➋ It takes away almost all possible angled shots your opponents can make and forces them to return the shot back to you down-the-middle;

- ➋ When both opponents are at the baseline, it usually results in both players moving to the center of the court, thus strategically opening up both alleys for your next angled shot; and,

- ➋ When both opponents are at net, they have little time to decide who will take the shot.

Comparatively, very few shots are hit down-the-line. So, what is the most important area of the court to be prepared to cover? Right! The middle. This not only applies to the net player who should

not be attached to "covering the line," (see section on SERVER'S PARTNER) but to BOTH partners, who must move together on the court in such a way that the middle is never left wide open.

Thus, when your partner is pulled off the court on a wide serve or wide shot, then it is your job to move to the center "T" until your partner has time to recover. AND, because you are the only one covering the entire court in such situations, it is incumbent upon your partner to return to her proper court position as soon as absolutely possible so you each can cover your assigned areas of the court. Similarly, when your partner is pulled to one side of the court (but not off the court), it is your obligation to move a few steps toward the center line to cover that area.

DEEP vs. SHORT

The question is always asked whether it is better to hit deep or short to your opponents. It is generally thought that the deeper the better as this puts your opponents under the most pressure, sometimes forces them to hit the ball while they are running back toward the fence and gives you more time to intercept the next shot (as the ball will be hit by the opponent farther away from the net).

But let me give you an alternate way of looking at this question: the doubles court is 78' long x 36' wide. If each opponent is responsible to cover half of the court, the most they will have to run from side to side is 18' (excluding a lob over the net player's head, in which case the maximum run is 36'). If, however, they have to run from 3' behind the baseline to a ball that lands 3' from the net, they have to travel 39' – more than twice the distance.

The resultant conclusion is that maybe it is better to hit short than deep – particularly against an opponent who is less than 100% mobile. It also provides a set up for your next shot to be a lob. Think about it. Try both and see what works.

"EASY" SHOTS

How many times have we hit a difficult shot only to miss the next "easy" shot? First, assume that no shot in tennis is "easy." A shot may appear "easier" in that we may not have to run for it, or it is a "sitter," or it was not hit hard. The problem is that our eyes light up and, without focusing on proper form, we try to blast a winner – all the wrong things to do.

A few suggestions:

- REPEAT: there are no "easy" shots;

- The "easier" the shot appears to be, make sure that you move your feet even more than you would for a hard shot. The "hard" shots automatically require us to move our feet – the easy shots don't – and so we don't. The main reason people miss easy shots is the failure to be in the proper position to hit the ball.

- The temptation is to run directly at the ball. Don't do it. Run to the side of the ball where you have the stronger side to hit from – for example, run to the left of the ball so you can hit a forehand.

- Don't take a big backswing – only a short one. Prepare as if you were going to volley the ball.

- Placement is more important than power;

- Use a volley stroke;

- For high sitters, do not swing at the ball and do not hit down at the ball. Hit "out." Your racquet face should end up being pointed toward the target when you finish your stroke. Most missed high shots go into the net because the racquet head is swung downward after the hit. So "freeze" your racquet after the shot to make sure the strings are pointing toward your target.

- Hit to safe targets (such as the targets used when volleying – see VOLLEYING section)

- Be prepared for the ball to come back

- RULE # 1: GET THE BALL OVER THE NET AND INTO YOUR OPPONENTS' COURT.

COMMUNICATION WITH YOUR PARTNER

Rule #1: NEVER criticize your partner during a match. In fact, never mention ANYTHING negative about your partner's play during a match. This includes non-verbal "statements" such as heavy sighing, throwing your hands up in the air, etc. The only discussions you should ever have need to be upbeat, encouraging and positive. You can (and should) have discussions about strategy – what is working and what is not; what are your opponents' strengths and weaknesses; what your opponents are doing to win points and what you can do to respond.

During a match is NEVER the time to give instructions to your partner about anything. There will be plenty of time to discuss all of that after the match.

Rule #2: ALWAYS call for the incoming lob by yelling "mine" or "yours" – and your partner responding "ok" or "I got it." Not only does it clarify who will be taking the ball, but it gives the person hitting the ball extra confidence by calling it out loud.

Rule #3: Always let your partner know when you are switching or want them to switch. The general rule of thumb is that the person furthest from the net makes that call as that person is the only one who can see the entire court.

Rule #4: If you have set plays, make sure you and your partner are clear about what is going to happen and what each partner's responsibilities are.

Rule #5: If your partner is down or is going through a tough stretch and not playing well, they need your support, confidence and assurances that everything is ok. Maybe taking a little extra time to have your partner tie their shoelaces or give them a breather is a good idea. Discuss this off the court so that you both know what works for each other.

ATTITUDE (INNER AND OUTER)

OUTER ATTITUDE

You cannot gain the upper psychological hand in a match by displaying any emotions other than positive ones. Every displayed negative emotion reduces your commitment to winning, interferes with your focus and emboldens your opponents. This means showing no

- Anger

- Frustration

- Weakness

- Complaining

- Dissension with your partner

- Signs of being tired

- Slouching body posture

- Glum facial expressions

- Making excuses

On the contrary, you should *always* appear to your opponents and your partner to be

- Energized (the way Nadal runs onto the court)

- Upbeat

- Confident

- Vocally celebrating winning points with your partner

- Openly and loudly congratulating your partner on great shots

And, if you really don't feel that way, then FAKE IT. You can fool your opponents and fool yourself into feeling that way.

INNER ATTITUDE

Your inner attitude not only affects your outer attitude but it controls the way you feel during the match. There are several tricks to use to improve your inner attitude:

- Distracting thoughts: one of the oldest and best ways of dealing with distractions (internal and external) while playing a point is the "Bounce/Hit" trick. It is very simple. Every time the ball bounces toward you or your opponent silently say to yourself "Bounce." And then say "Hit" when the ball is hit. It preoccupies your mind with something about the game to focus on other than distracting thoughts.

- Breathing. When tension arises, use the thousands of years old meditation and yoga trick of breathing. Between points, just focus on your breathing. Focus on each breath between points and before serving. Say to yourself: "breathe in; breathe out." When you are feeling particularly overwhelmed and have the time during cross-overs or between sets, try this: breathe in through your nose deeply and slowly to a count of 4; hold your breath for a count of 4-6; and then slowly breathe out through

your mouth to a count of 8. Do this 3 times in a row. It works – on and off the court.

- The "black box." Gigi Fernandez, one of the most winning doubles players in history, used to be known as the John McEnroe of women's tennis. She had to work hard at changing her attitude – both internal and external. She says that her most successful trick was her invention of the "black box." She merely imagines a black box (with a latch) in her mind. Whenever any negative thought, negative emotion or distracting thought enters her mind during a match, she puts it in the black box, latches it and waits until after the match to open it. She says it has changed her attitude completely and is also her most popular trick with her students.

- Confidence. If you don't have it, work on it. Confidence can be learned and made a habit. Even the Bryan Brothers used a technique. Their father, noted tennis coach Wayne Bryan, created a half-hour video of his sons in televised matches, recording only the most spectacular shots and wins, along with the announcers' amazed comments. The Bryans watched that video before every match they played.

- Differentiate between what you can control in the match and what you cannot control. Focus on what you can control and let go of what you cannot control.

- BETWEEN POINTS: 80%-90% of your time on the court is between points. What you do with this time is important. Dr. Jim Loehr specializes in tennis psychology and has developed a "16-second cure." Between each and every point you go through a 4-step process. You will see that all pros do this or something very similar. He recommends this process to achieve stabilization between points, each step of which lasts 4-5 seconds:

- A positive response to the point that has just ended. Be physically and psychologically "up"; do not complain or berate yourself;

- Relaxation – generally through breathing (see above).

- Evaluate what is happening on the court and plan your next point;

- Engage in a ritual or routine (like Nadal), which acts as a safety mechanism for the brain, deals with the stress and tells your brain that it is time to focus on the next point.

- Read:

  - "The Best Tennis of Your Life," by Jeff Greenwald – the leading expert on tennis psychology;

  - "The Champion's Mind," by Jim Afremow, PhD;

  - "Mindset," by Dr. Carol S. Dweck;

  - and the old standard: "The Inner Game of Tennis," by W. Timothy Gallwey.

- Lose your ego. Do not define yourself as a human being by whether you win or lose a match. It puts far too much pressure on yourself.

- Learn to see success and happiness in learning and getting better – without having to win.

- If you are winning, do not converse with your opponents during cross-overs. You will lose your focus.

- Meditate. The entire purpose of meditation is to learn how to accept your thoughts and let them pass without becoming attached to them. It not only improves your life, but your tennis mind. Try starting with just 10 minutes a day with the App "CALM." Read my favorite book of all time by Jon Kabat-Zinn, "Wherever You Go, There You Are." I have read it 7 times. "Meditation is a lifelong gift. It is something you can call on at any time." Paul McCartney

- REMEMBER: 85% of the time on the court you are not playing a point – but are between points. Learn how to use this time wisely and not let your mind defeat you.

USTA RULES AND THE CODE

YES, there ARE rules promulgated by the ITF and followed by the USTA. And there is a Rule Book. BUY ONE! Make sure it includes the supplemental CODE. Read it. Study it. Mark those Rules mentioned in this book and all others you want to have at your fingertips during a match. Tab those pages with a marked tab so you can quickly locate the ones you need right away. You will not only win all on-court arguments about rule-related issues, but you will be able to support many of the tactics outlined in this book that some players may challenge you as being illegal. And, as important, your preparedness and knowledge of the Rules will by itself intimidate your opponents.

These Rules govern all USTA matches and can be used by you as the only written rules for all matches.

FOR INSTANCE DID YOU KNOW THAT –

- The returner can stand anywhere on her side of the net – inside or outside of the lines. USTA Rule 8

- The returner's partner may stand anywhere on his side of the net – inside or outside of the lines. USTA Rule 26

- The server's partner may stand anywhere on his side of the net – inside or outside of the lines. USTA Rule 26

- Underhand serves are perfectly legal, as long as the ball is "released" by hand before it is hit by the racquet. USTA Rule 16

- If a player's cell phone rings during a point, the opponent may stop the point and claim the point based upon a "deliberate hindrance." There is no penalty for a vibrating phone. USTA Rule 26, Comment 26.3. Make sure it is not your partner's phone before invoking this rule!

- A player may change position *at any time* – including when the server is tossing the ball. The CODE, Principle #35.

- On clay or HarTru courts, a player is not required to show the opponent any mark. The CODE, Principle #21

- On clay or HarTru courts, a player is not allowed to go to the other side of the net to inspect a mark. The CODE, Principle #21. In fact, doing so is considered Unsportsmanlike Conduct.

- If one partner calls the ball "out" and the other partner sees the ball as being "in," the ball is "in." The Code, Principle #14

- If a ball cannot be called out "with certainty," it is good. The CODE, Principle #8.

- A player is entitled to win a point if he has been "hindered" by the opponent's "intentional" action - that is, if the opponent intended to do the "act" that hindered. It is not necessary for the opponent to have intended to hinder the other player, merely that she intended to perform the act which caused the opponent to become hindered. USTA Rule 26

- The server shall not serve until the returner is "ready." However, if the returner "attempts to return the ball" he "shall be considered as being ready." USTA Rule 21. If not ready, the appropriate procedure is to put your hand up in the "STOP" position until ready to proceed.

- The server is allowed to stand behind his own alley. USTA Rule 16: "...the server shall stand with both feet behind...the baseline and within the imaginary extensions of the centre mark and the sideline."

- After a 7-point tie breaker, the players change sides. USTA Rule 10, Comment 10.1

- Who serves after a tiebreak? "The player/team whose turn it was to serve first in the tiebreak game shall be the receiver in the first game of the following set." USTA Rule 5.

GET THE BALL OVER THE NET

How basic can I get? It may sound stupid but this goal is not always first on everyone's mind. Rather, players tend to think of great winning shots and impressive ways to win points – not all of which are high percentage shots.

PRACTICE hitting your shots with lots of net clearance (and side-line clearance by aiming for inside the singles lines). Use topspin to give yourself more net clearance, if necessary. Hit deep lobs. You can't win the point if you hit the net. But you CAN win the point by hitting a bad shot that goes over the net and gives your opponents the chance to miss their next shot. AND RUN FOR EVERY SHOT to get it back. It is amazing how many times you can win a point by getting the ball back over the net just one more time.

A NOTE ON STRETCHING

According to physiologists, as you age your muscles shorten. Also, the more you exercise, the more your muscles shorten.

INESCAPABLE CONCLUSION? As you get older and continue to play lots of tennis, it is imperative that you stretch – regularly.

GAME-TIME RX

Having issues during a match? Missing shots? Not playing your best? Make a copy or take a photo of this page and keep it with you to read during change-overs. (there is a second identical page at the end of this book that you can remove to put in your tennis bag).

GENERAL TIPS

- FOCUS ON THE BALL; focus on a small part of the ball or on the color on the ball;
- Concentrate only on moving your feet;
- Use "bounce-hit" method explained in ATTITUDE section

SERVE PROBLEMS

- KEEP THE TOSS HIGH; don't grip the ball tightly
- Keep chin up and left arm up as long as possible

VOLLEY PROBLEMS

- KEEP RACQUET UP AND AS FAR IN FRONT AS YOU CAN KEEP IT
- Move feet toward ball
- "FREEZE" your racquet after the shot to insure that the racquet face points toward target
- Aim for targets vs. using power

GROUND STROKES

- KEEP KNEES BENT

- KEEP HEAD DOWN UNTIL SHOT COMPLETED

- Step into the ball

- Focus on lots of net clearance.

OVERHEAD

- MOVE FEET – A LOT

- Stand side-ways

- Keep chin up and left index finger pointing at ball until contact

- Aim for just behind the center "T"

ATTITUDE

- Slow down between points; take time; re-tie your shoe laces

- BREATHE (see ATTITUDE section)

- Act confident and "up"

- smile

POSITIONS

THE SERVER

GENERAL PRINCIPLES

RULE #1: HOLD SERVE! If your team is able to hold serve, then you have a huge probability of winning a match as all you have to do is get one break per set on your opponents' service games or win a tie-breaker. It sounds easy and simple, but it is a truism that should be kept in mind when serving.

RULE #2: GET 75% OF YOUR FIRST SERVES IN.

RULE #3: WIN THE FIRST POINT. According to statistics, the pros hold serve 93% of the time if they win the first point.

CONTROL

The server is in the unique position of being the only player that can control the tempo of the game. Use that control strategically - particularly when the match momentum seems to start going against you. Take your time. Bounce the ball a few extra times. Novak's record is 32 bounces.

CONSISTENCY

The general rule of thumb is that you should make 75% of your first serves in doubles. You want to get your first serve in - not just because it is a better serve, but because the returner is psychologically "up" for a second serve and is thinking much more aggressively. If you need to, reduce your power by 10%-15% to make your first serve more reliable. Getting your first serve in when playing doubles is more important than when playing singles. Developing and sticking by a service ritual will help consistency. See below.

DEVELOP AND USE A RITUAL

Think Rafa Nadal, but not to the same extent! While he is an extreme example, he nonetheless demonstrates the importance of a pre-serve ritual to calm the nerves and focus the mind. Always bounce the ball the same number of times; breathe deeply.

SPEED vs. DEPTH vs. SPIN vs. LOCATION

It is almost universally accepted that you should sacrifice speed for depth, placement and spin - in that order of importance. So, when you practice, focus on those factors in that order - with depth being the most important of all.

VARY THE SERVE

You may think that you only have one serve. Think again. You already have at least two different speeds of serve that you are using for your first and second serves (actually, you are capable of serving at an infinite number of different speeds). Hopefully, you also have a slice serve and, if you are very fortunate and have worked very hard, you also have a topspin serve. At the very least, you have 3 different locations to serve to: wide; down the center "T"; and at the body. Practice all three locations (see PRACTICE section). BUT you can vary your serve even more by serving from at least three different points on the baseline: the alley, the center line and, the standard location, half-way between those two points. You are allowed to stand on the baseline behind the alley to serve. See USTA Rule 16: "within the imaginary extensions of the centre mark and the sideline." So now you have at least three different serve targets from three different baseline locations at two different speeds. That means that you have at least **18** different serves to both the deuce and AD courts. And, if you have a slice to add to the equation, you really now have at least **36** different serves to each side of the court. Adding a topspin serve gives you a ridiculous number of at least **72** different serves to each court. And that's without adding a "Junk" serve to your repertoire. The secret here is to move your starting position on the baseline. This not only gives you tremendous variety, but gives the returner a different look on almost every serve, makes them wonder what you are up to, gives you more slice wide and more slice into the body in the deuce court from the alley and prevents the returner from getting into a rhythm.

UNDERHAND SERVE

It maybe "underhanded," but the underhand serve is not illegal as long as the ball is not in your hand when you make contact with the ball. The ball does not need to be "tossed" in the air, but merely "released by hand in any direction first." USTA Rule 16. This serve is certainly a surprise to the returner, especially if used as a first serve when the returner is standing further back behind the baseline and will not be able to reach a short spinning serve. The trick is to hit the ball short with as much side-spin and under-spin as you can put on it at the same time, causing the ball to drop just over the net and jump dramatically to the returner's left. It is especially effective when serving to the AD court as the ball ends up off the court. A good starting point to learn this serve is to watch the YouTube demonstration by Gino Fava.

You may not make friends but you will win points.

TARGETS

DOWN-THE-MIDDLE: this is what is considered to be the standard location for about 75-85% of your serves. The reasons:

- ❧ it takes away any angled return that the returner can make and gives your net partner the automatic

chance to pinch to the middle of the court, making sure that they "mirror" the ball. This (1) gives the net person a much greater opportunity to pick off a return, (2) puts more pressure on the returner and (3) squeezes the space that the returner has to aim for.

- You must, however, take into account what the returner's strengths and weaknesses are. If the returner has a weak backhand, then, when serving to the AD court, it may be better to serve to that weaker side, adding in down-the -middle serves for variety. Finally, if serving down-the-middle to the AD court, it is wise to add slice to the serve so that it spins away from the returner.

- By serving down-the-middle, you are serving over the lowest part of the net, decreasing your chances of hitting the serve into the net.

- TACTIC: your net partner should ALWAYS jump to the center line of the court when you serve down the middle unless the returner has an incredible return down the line from the middle of the court. Make sure you and your partner discuss and agree upon this tactic before the match. It should become automatic.

WIDE: Adding spin to this serve makes it much more effective (slice to the deuce court and topspin to the AD court). With spin, this serve can take the returner off the court and/or reach for the ball resulting in a weak return. The downsides? If you miss by just a little, the wide serve to the deuce court goes to the returner's forehand wheelhouse for a powerful return. Also, when you give the returner a big angle, you also give her the chance to return with a big angle.

<u>THE BODY SERVE</u>: For some reason, most players practice the wide serve and the down-the-middle serve – but not the body serve. It is an extremely effective serve – particularly against returners who have long arms, who are not very mobile or who have a two-handed backhand (aim at their right hip). Learn it and practice it. It is most effective when *sliced* into the body. A good body serve will usually result in a weak return. Because you are aiming for the middle of the service box, it is also the safest serve, least likely to go wide.

THE AUSTRALIAN FORMATION

I assume for this book that you already know this formation. If not, it is ridiculously easy to learn – but it must be practiced so that both partners feel very comfortable using it. It has many benefits:

- ❧ ELIMINATE CROSS-COURT RETURNS: The best reason to use this formation is to counter a returner who hits very effective cross-court returns – either with angles, slice or power. By placing the net partner directly in the path of that type return, it forces the returner to use another type of return. Many players (in particular the ones who have fabulous cross-court returns) become married to their return because it is so successful. In many instances, just using that formation forces your opponent to give up their best weapon and draws errors from him – sometimes multiple errors on consecutive returns – and has the added psychological benefit of taking away their favorite shot and frustrating them in their inability to deal with the change.

- ❧ GIVE YOUR OPPONENTS A DIFFERENT LOOK: Many times, your opponents will not be very familiar with this formation and it will confuse them. Even if they are familiar with it, it is a different look

and should be tried any time you are having difficulty holding serve.

- BETTER SERVICE POSITION: As you, the server,
are serving from the center line (and usually serving down-the-middle) you are serving over the net
strap – the lowest part of the net – which reduces
the number of serves into the net.

- FOREHAND ADVANTAGE vs. BACKHAND ON AD
SERVES: Because the AD court returner will usually respond by returning the ball down-the-line to
the open court, when you serve to the AD court,
the returner's shot will be going to your forehand
instead of your backhand. And, if you end up in a
rally, it will be one between your forehand and your
opponent's backhand, giving you the advantage.
More than one pro recommends using this formation *most of the time* when serving to the AD court
for this reason alone.

- SERVE AND VOLLEY: This formation still allows the
server to serve and volley – and, as with ground
strokes, you are most likely to get a return to your
forehand to volley.

- POACHING: The net person can still poach from
this formation and probably should do so on occasion if the formation is employed often.

- THE "AUSTRALIAN JUMP": This involves the server
serving from the center of the court (as in the
Australian formation), but the net person assuming the standard non-Australian net position to
start with. The net person peeks at the server and,
just when the server tosses the ball, the net person jumps into the Australian net position. This can
cause confusion for the returner and may result

in the returner having to immediately rethink and change their planned return.

● BOTH BACK AUSTRALIAN FORMATION: Just to introduce some more confusion to the opponents' minds, you may want to throw in this variation of the Australian formation on some occasions. The server's partner (who usually plays the net position) plays back on the baseline – so that both server and partner are standing at the baseline on the same side of the center line. This is a highly unusual formation and can be confusing to your opponents. As an additional variation, the non-serving "net" partner can charge straight ahead to his normal Australian position after the ball is served into the returner's box. If you choose this latter tactic, make sure the server gets the first serve in, as having to repeat this tactic on a second serve loses its surprise value after your opponents have just seen it.

● BOTH BACK "AUSTRALIAN JUMP": either (1) start in the "Both Back" Australian formation and, as your partner serves the ball, jump into the standard Australian formation; or, (2) start in the standard Australian formation and, as your partner serves, sump into the "Both Back" Australian Formation.

FIXES

While this book is not focused on stroke technique, here are some fixes for the serve that are so common and so universal that I feel it necessary to include them in the event you encounter problems.

THE SERVE INTO THE NET: Most serve problems involve serving the ball short into the net. Check out the following fixes for this problem:

- Toss the ball higher;

- Keep your head up *and* your chin up until contact is made;

- Keep your tossing arm up for a count of "one thousand one;"

- Make sure you reach up to hit the ball;

- Try practicing serving the ball so it reaches the opposite fence. This will groove your serve to hit deeper.

ERRANT TOSSES: Sometimes people get into a pattern of the "yips" and cannot find the right toss. This is usually caused by bending your elbow during the toss, causing the ball to go in different directions. Focus on locking your elbow and setting your wrist so that the palm faces upward. If necessary, wear a brace (like the ones for tennis elbow) over your elbow to remind you to do this. Bad tosses can also be caused by gripping the ball too tightly. Relax, or use the palm of your hand to toss the ball rather than your fingers.

THE RETURNER

Pros differ as to whether this is the most important or second most important position. All pros agree, however that the first three rules for the Returner are:

1. GET THE BALL BACK;

2. GET THE BALL BACK; and,

3. GET THE BALL BACK.

Statistics show that the team with the most consistent returns of serve wins 90% of matches. Statistics also show that more shots are missed on the return of serve than any other shot. Why? First:

no one ever practices this shot; Second: when practicing, you need someone to serve to you and, preferably, someone to pose as an aggressive net player to make the return situation more realistic and more pressured.

Here are some fundamental "return of serve" techniques and tactics:

STARTING POSITION

While you probably have your own favorite starting position in returning, it is a good idea to start two steps behind where you want to end up hitting the ball. Start in a crouch, with knees bent and feet wide apart so you are (and appear to be) ready and hungry. You are more intimidating to the server in that position. Watch Novak Djokovic. When the ball is tossed, take two steps forward and split step on your toes as the server hits the ball. This process will make you the most prepared you can be to move from side-to-side to return the ball.

MIX UP YOUR STARTING POSITIONS

Start each point by standing in a different place. It gives the server different looks and makes them think about what you are trying to do. You can vary this and be more extreme about it for the second serve than the first serve, although you can experiment with it on the first serve as well. Try standing in the alley, or right at the center line. You can even stand outside of the alley. This makes the server try to hit the serve in the more open space. That may not be the server's favorite spot to hit – or it may be the spot you want him to hit if it is toward your strength. In either event, as soon as the ball is tossed, you can move toward the center of the box, putting yourself in your best return position. It is legal to stand anywhere - even outside of the alley. See, USTA Rule 8; Case 1, Decision:

"The receiver may take any position inside or outside the lines on the receiver's side of the net."

and it is legal to change positions while the server is tossing the ball. USTA Code Principle #35:

"A player may change position at any time, including while the server is tossing the ball."

CUTTING OFF THE BALL

Unless the ball is right in your swing zone, you will have to move to the right or left. Do so at a 45% angle so that you are also partly moving toward the net at the same time you are moving laterally. This is the shortest distance to the ball and will provide at least some forward momentum toward the net when you hit the ball.

RETURN TARGETS AND TYPES OF RETURNS

Try to make every return a difficult shot for the server (or the server's partner) so the shot will set up your net partner for a put-away or you for an easy first volley from the service line. There are more types of returns than most people think. NEVER return a simple shot to the service line to the server's forehand, unless it is very wide and sliced. Such a return will instantly give the server the advantage of going on the offense with a variety of easy shots. Try to make as many returns as possible as an approach shot (see section on APPROACH SHOTS).

Good returns include:

- Deep high-bouncing topspin to the server's backhand
- Deep slice at the server's feet or to their backhand

- Slice or hard flat shot down the middle – especially effective if the server's net partner never poaches

- Deep high-bouncing lob down the middle to the server's backhand

- A deep lob over the net person (over their backhand side to make it more difficult for the net person to hit the ball well if the lob is short)

  NOTE: After making this return and the ball going over the net person, your net partner and you should assume the "I" formation, with your net partner at net or the service line covering the short return and you covering the deep lob – or, in the alternative, switching with your partner so the net person is directly in front of the ball and you covering the other side of the court.

  NOTE: If the serving team responds to future lob returns by starting with the net person backing up to the service line, continue to lob over that person's backhand – still a very difficult and awkward shot that will probably result in a weak reply. For some unknown reason, when the net person starts at the service line almost all players give up trying to hit the lob over that person. In the alternative, slice the return low at the opponent's feet, just to the left or right of them.

  NOTE: It is best to let your partner know that you are planning on making this return so that they can be prepared to move to the net.

- A short slice down the middle, forcing the server to move forward and hit up, giving your net partner a chance to pick it off

- Short angled return to the side "T"

- Down the line – especially to a weak net player and especially to the backhand from the deuce court. See SET PLAY at p. __.

- Drop shot P. 49

- At the feet of the net person if they are standing back at the service line to protect against the lob

- Chip-and-Charge: use any of these returns, but stand closer to the service line (much easier on a second serve), take the ball on the rise with a slice return and move into position inside the service line. This puts tremendous pressure on the server and places you on the offense.

- RETURNS AGAINST THE AUSTRALIAN AND "I" FORMATIONS:

  - The first temptation is to hit down the line to the open court. This is effective, particularly by hitting a deep slice down the alley; but it is not always an easy shot, as you are hitting over the highest part of the net to a narrow target.

  - The best return is a lob over the net person. It is directed toward the deepest part of the court (giving you 4 1/2' more of court to work with). As the server will be moving to the other side of the court to cover the down-the-line shot, it will be difficult for them to reverse direction to cover the lob. It is best for the AD court player to make this return as it goes to the server's backhand side.

- ○ Also good: a drop shot to the open court, although the net person in the "I" formation may be moving in that direction

- ○ Also good: a shot right at the net person in the "I" formation, as the net person may move in one direction or the other and the shot will go to where they were – not where they are heading to.

TACTIC: There are 9 different returns listed above. Use them all and mix them up so that your opponents do not know what to expect and do not get into a pattern. Pay attention to which ones work the best so you can go to them on big points.

TACTIC: WIDE SERVES. When the serve takes you wide off the court so that you have to run and reach, your entire court is left exposed with your partner being the only one to defend it. You do not want to hit a return directly to either opponent. So, the most effective return is a lob down-the-line as (1) it puts the ball over the head of the net person for a difficult shot and (2) it gives you and your partner time to recover your positions.

TACTIC: As an alternative to constant variation, you can return 3 times by lobbing over the net person and then drop-shotting cross court.

TACTIC: Establish your down-the-line return early in the match to freeze your net opponent from poaching. It is amazing how one shot down-the-line can discourage a player from poaching for an entire match.

TACTIC: As just illustrated, establish a pattern and then unexpectedly change it. As an example, make most returns deep cross-court from the AD side to the server's backhand. They will start to expect this return and move to their left after the serve to prepare for it. If you then hit a hard slice down the middle, the server will not be there to get it.

ESSENTIAL TACTIC: Move into the ball as you hit these returns and follow all good returns (ones that put the opponents on the defensive) to the service line for a follow-up volley, maintaining maximum pressure.

MUST DO: If you return with a drop shot, it is imperative for your partner to charge the ball and "smother" it to intimidate your opponent and to take away all angles the opponent has.

PRESSURE TACTIC - SQUEEZING

Should the opponent have a weak second serve or lack confidence in it, try standing at or just behind the service line. This places tremendous pressure on the server and challenges the server's ego – particularly when the server is male and even more so if the returner is female. As added pressure, while doing so, you can have your partner stand right at the corner of the service line and center line, thereby seeming to squeeze the serving area of the box into a smaller area. This also gives you the advantage of being able to take a weak second serve and chip a deep slice return and attack the net. Finally, a return from the service line takes time away from the server in hitting your return and takes time away from the net person should you choose to return the serve right at the net person. For a while, Federer even tried this technique of standing just at the service line and half-volleying or chipping second returns. Totally unexpected.

TACTIC: MOVING DURING THE TOSS

This is entirely legal and specifically allowed by USTA CODE #35, which provides:

> "A player may change position at any time, including when the server is tossing the ball."

So, you are entitled to start in one position, and as the toss is being made, move to another position.

TACTIC: CHANGING POSITIONS

Notice how many times the server looks up at the service box before beginning the service motion. If it is more than once, note this fact and change positions after the first look so that you are standing in a different location for each look. This will cause the server to change the direction of the serve at the last minute, increasing the likelihood of mistakes.

TACTIC: "CREEPING"

As it is legal to move anywhere any time on the court, it is legal to slowly "creep" forward from your starting position toward the service line while the server is focusing on where to serve the ball. When the server sees you doing this, it adds enormous pressure on him and, if the server focuses too much on you, then he may double fault.

TACTIC: SWITCHING

It is very effective for you as the returner to switch positions with your partner after you have returned the serve - particularly if the return is deep. It confuses your opponents and can be used to place the stronger volleyer on your team in the position where the server usually will hit her next shot expecting it to go to the weaker partner. This is especially effective when the return is a very deep high-bouncing shot, which will induce the server to hit a lob, thereby placing the partner with the stronger overhead close to the net.

THE SERVER'S PARTNER

This is the most FUN position on the court and the position where you are able to most affect the outcome of the point.

PURPOSE: To impact *every* point that your partner serves!

As the net player you have several options as to what to do as your partner serves and as the opponent returns. <u>STANDING IN ONE PLACE IS NOT AN OPTION</u>.

Your options are:

1. POACH: see section on "POACHING". If you have any fear of poaching or of "getting beaten down the line," get over it. You will be a significantly better player and have much more fun by losing that fear. An active net player is threatening, intimidating, distracting and downright disturbing to the returner. Active poaching causes the returner to keep from getting into a groove of cross-court returns. It causes the returner to think about you rather than focusing on the ball. Your constant movement at net is a distraction. It gives you the chance to put the ball away for a winner in one shot. It gives you and your partner a feeling of domination and destroys your opponents' confidence. If you never poach, you might as well stay at the baseline or play singles.

   IMPORTANT NOTE: <u>DO NOT "COVER THE ALLEY"</u> - COVER THE MIDDLE

   Many players tell their partners to "cover the alley" because they are afraid that returner will hit the return there. Have a chat with you partner about this. Hitting a return down the alley (particularly from a ball served down the center "T") is one of the hardest shots in tennis. It is a very narrow area in the alley to aim for and must travel over the highest part of the net. In addition, if you stand one step toward the alley from the middle of the return box, you will notice that with two more steps you can reach all returns "down the alley" other than the perfect one that lands on the line; and with two steps in the other direction, you can cover most shots toward the middle. Make your opponent prove that they can hit a down-the-line winner – more than once. The adage is that, if you are not getting passed down the line once or twice in a set, you are not poaching enough or are starting too close to the alley to begin with. On the other hand, if your opponent is capable of repeating down-the-line winners, then you must respect that shot and be more prepared to get to that return.

In general, it is much more important to **cover the middle** of the court than the alleys. See "COVER THE MIDDLE" section above. The middle of the court is where most winners are hit. Remember the old saying: "down-the-middle solves the riddle." It is the most important part of the court to always cover. Especially when your partner must move to their alley or off the court, your obligation is to move to the center "T" to cover you part of the court and your partner's part of the court.

TACTIC: Should the returner pass you down the line, on the next serve to that returner move very close to the singles line to make the returner think you are afraid he will hit that shot again. Then poach.

2. FAKE POACH: see section "POACHING."

3. MIDDLE-JUMP: This is a concept invented by the Bryan Brothers. It involves a "half poach" – poaching

only to the center line to see if you can pick off the ball. When you get to the center line, if you can pick off the ball by going further, then continue to where you need to go to hit the ball. If not, then retreat to your side of the court. Your partner does not switch unless and until you cross the center line.

4. PINCH: Again, this concept was used a lot by the Bryan Brothers. It involves moving partially toward the net strap - without actually poaching or middle-jumping - but squeezing the space that your opponent has to hit into.

5. MIRROR THE BALL: A very simple and effective way of always staying in the middle of the possible angles the returner can hit into is to always stay in front of the ball ("mirroring it") when it is on your half of the court. Thus, if your partner serves down the center "T," then move (diagonally) toward the middle. This does not apply to balls that are very wide where your opponent is in the alley or outside of the court.

6. ALWAYS BE MOVING: To start, take a position in the middle of the box, but closer to the service line. When you hear your partner serve the ball, take two steps toward the net and split-step. From that position, you are balanced, on your toes and ready to move in any direction.

TACTIC: Use different starting positions. In addition to the Australian or "I" formations, you can use the "Australian Both Back" formation or both back at the baseline to give the returner different looks. For some reason, seeing both the server and his partner at the baseline can be confusing to returners.

You should also be willing to change your position based upon the returner's habits. If the returner likes to lob over your head, start just in back of the service line. Every player will stop lobbing you - although, as discussed in the RETURNER section, there is no reason to stop. But it works. And, you can poach from the service line too – which is effective because no one ever expects you to do it. Also, if the returner is successfully pinning your partner to the baseline and avoiding you at net by hitting high lobs back to your partner, start at the center "T" to make them think about aiming their return elsewhere.

TACTIC: SECOND SHOT POACH. Returners are very alert to a possible poach by the server's partner on the return of serve – particularly when the net person has proven themselves to be active at net. What most returners do not expect, however, is for the net person to poach the returner's second shot. Try it. It's fun and very successful.

DON'T FORGET: it is usually the net person's responsibility to chase down drop shots

IMPORTANT: Keep moving! Before the serve, during the serve, during the return and throughout the point. Your obligation to impact the point does not end when the return is made back to your partner.

THE RETURNER'S PARTNER

This is the most underappreciated and underutilized position on the court – but can and should have a significant impact on every point. How to impact every point most effectively?

YOUR FIRST OBLIGATION

Before employing the tactics mentioned below, it is important that the returner's partner call any serves that are long. See USTA CODE, Principle #25:

"In doubles, the receiver's partner should call the service line, and the receiver should call the side-line and center service line. Nonetheless, either partner may call a ball that either sees clearly."

You have by far the best view of the service line as you are usually starting from a position directly next to the returner's service line.

YOUR SECOND OBLIGATION

Protect against the poach by the server's partner. To do this effectively, after you have observed the serve being "good", do <u>not</u> look back at your partner to see what kind of return is being made. You will be able to know where the return is going based on the movement of your opponents. Instead, you must focus immediately and only on the opposing net player, as that person is the immediate threat to you. Always stand facing the opposing net player on your toes and with your racquet up. ALWAYS be prepared to return a poached ball until the ball has passed the net person, at which time you can focus on the server.

VARY YOUR STARTING POSITION

Learn to stand in a different place for every serve and learn to move in different directions. This gives the server a different look each time. Some examples:

- If your opponent has an unreliable serve, put pressure on her by "squeezing" the server - moving to a position where you are straddling the corner of the service box. This puts pressure on any server – but particularly one without 100% confidence in their serve. It forces the server to try to serve to a visually much smaller area in the service box and forces many mistakes.

- LEGALITY: some players object that doing so is illegal. Pull out your trusty marked-up Rule Book to USTA RULE 26, Case #5:

> *"The server's partner and the receiver's partner may take any position on their own side of the net, inside or outside of the court."*

So, hypothetically, both the returner and her partner can stand side-by-side behind the service line on the same side of the court behind the service box – although only the actual designated returner may return the ball.

TACTIC: if this "squeezing" position is taken by the returner's partner *and* combined with the returner's tactic of moving forward on the second serve to stand on or just behind the service line (or in the alley) it places extraordinary pressure on the server and results in many double faults. If the server beats you by serving an ace, just say "Nice serve" and try the tactic again to see if the server can perform this feat twice in a row. Do not assume that, because they succeeded once, they can do it again.

NOTE: To combat this tactic, the server can aim the serve at the returner's partner and, if the ball hits him in the air, the server wins the point. See, USTA Rule 24 (i): "the point is lost if the ball in play touches the player or anything that the player is wearing or carrying, except the racquet. . . ."

POACHING

There are several scenarios in which the returner's partner can successfully poach and/or impact the server's next shot.

- If your returner/partner can hit a good low return against a serve and volleyer, then look for the server's racquet to dip below the level of the net for a low volley or half-volley. As soon as you see this, poach (at a diagonal) toward the net strap, ready to pick off a weak shot that the server has to hit up to get over the net.

- Server's first ground stroke: no one expects the returner's partner to poach on the server's first shot after the serve - whether it is ground stroke or a volley. Go for it.

- **SET PLAY:** Decide with your partner who receives in the deuce court that, if the serve goes to his forehand, your partner will return down the line to a right-handed net person's backhand. You will stand at the center "T". If it is to her forehand, as soon as your partner hits the ball immediately charge directly at the net person, making sure to keep your racquet up and ready for the next shot - which, according statistics, will be a weak reply toward the middle of the court. If so, the ball should go directly to you and you will be in a perfect position to put the ball away.

- **SET PLAY:** If you are playing the AD court, agree before the point that the return from the deuce court will be a lob down the line over the net person's head, making the server run to the opposite side of the court. After the lob passes the net person's head, move to close on the net - either on the middle service line or directly facing the location of the ball, ready to put away any weak shot that comes back from the baseline player. Your partner will remain back toward the baseline to cover any deep lobs by (1) standing directly behind you if you are in the middle of the court or (2) stand-ing toward the AD side, if you have moved over to the deuce side. These changes in positions of your team, in addition to being effective aggressive tac-tics, can also distract your opponents.

This same play can also be run from the AD court return, although (1) the lob will probably be over the net person's forehand and (2) will head to the server's forehand instead of their backhand.

- **SET PLAY:** The above play can also be effective at any time during a point so long as you and your partner are on the same page about what to do after a good lob.

- **SET PLAY:** On occasion it is fun and effective to switch sides after the returner has returned the ball to the baseline player. The server takes over your position on the service line, and you move over to the service line on the side the server had served from. This allows a stronger player to take a shot that was intended for the weaker player on the team. It can also be confusing and distracting to your opponents.

- **MUST DO:** if your partner returns with a drop shot, it is imperative that you charge toward the ball, dis-tracting and putting pressure on the opponent and cutting off the angles they have.

- **REMEMBER:** as the net person, it is usually your responsibility to chase down all drop shots made by your opponents.

SHOTS

THE LOB

The lob is the most misunderstood, underutilized and berated shot in tennis and is presumed by many in the game to be used only by players who can't hit a "real shot" or who are lob kings and queens. Because the most important and effective doubles strategy is to take over the net and to keep your opponents away from the net, the lob is really the most important and most effective shot in tennis. Learn it well and use it often!

Defensively, it puts the ball high and deep to your opponents' backcourt when you are out of position to (1) give your opponents a difficult shot, which many people miss, (2) get your opponents off the net and (3) give you and your partner time to recover to your proper court positions.

Offensively, on a return of serve or during a point, it can immediately place you on the offensive and your opponents on the defensive by forcing them back to the baseline while allowing you to advance to the service line or the net.

FORM TIP: Bend your knees and step into the ball when hitting a lob – just like any other shot. Bending your knees to hit a lob is important in that doing so will give more lift to the ball for a higher and deeper lob.

TIP: A slice lob gives you more control over the lob than a flat or topspin lob because the ball stays on the racquet strings longer.

TACTICS:

- ❧ On return of serve, lob over the net person's backhand side. If your lob is short, the net person will still have to hit a backhand lob. This is particularly effective from the deuce court as it forces the backcourt person to hit a backhand from deep in the court;

- Martina Navratilova recommends an AD court player returning against an active poacher to hit a medium-high lob down the middle so that the poaching net player can just reach the lob with their backhand, but not be able to do anything other than hit a weak shot.

- A lob return of serve deep to the server, if they are not serving and volleying, will keep the server behind the service line and will give them a high-bouncing shot to deal with.

- Most pros think that height is more important than depth. Give it a try.

- Whenever your opponents (or one of them) gets close to the net during a point, lob them. Be aware of how close to the net your opponents like to play during points. Many people like to move closer to the net during a point. If they do, LOB!

- Learn the Lob Volley. Hitting this shot will give you a winner and will take the air out of your opponents.

- If you have the time to practice, learn a lob volley on your "serve and volley" first volley.

- And then, the old standby: dropshot and then lob. A real killer.

- Crosscourt or down the line? Generally, you want to lob over the head of the person who is closer to the net. When your opponents are about the same distance from the net, always lob cross-court, as there is much more room for the ball to land. In fact, because you are hitting diagonally, there is an additional 4 1/2 feet of court from you to the far corner than from you straight to the baseline ahead of you.

- AFTER THE LOB: particularly on an offensive lob, once you know that the ball is deep enough to get over the net person, immediately either (1) both you and your partner go to just behind the service line and prepare for a weak reply or a short lob or (2) have one partner move up inside the service line and the other stand near the baseline. While you should be prepared to put away a weak shot you also need to expect that the return shot will be another lob.

TACTIC: THE REACHABLE LOB

Most players assume that that a lob, to be a good one, must pass out of the reach of the net player. However, it is also a wonderful shot if hit just within the reach of the net player and over their backhand side, so that they can hit it but do nothing other than hit a weak shot. Try it. It works.

TIP:

Some pros preach to never hit the lob long; i.e., it is better to hit a weak short lob that will keep you in the point than to lose the point by hitting it long. Whether that is a good philosophy or not, it is good practice to forget about trying to hit the "perfect" lob. If you push your opponent off the net and force them to hit a difficult shot, you have succeeded.

THE VOLLEY

REMINDERS

- Racquet head up

- Arms in front outstretched

- On your toes

- Split-step as the ball is hit by your opponent

- As the ball comes toward you –

  - Turn to the side

  - Hold your racquet throat with your left hand and only bring it back as far as you can without letting go with your left hand

  - Do not let the racquet get behind you

  - Release left hand only when the forward motion of the racquet begins. This means NO BACKSWING!

- Step into the ball ("volley with your feet")

- Do not swing at the ball – make a short punching motion toward the ball so that the face of the racquet ends up pointing toward your target

- Placement is more important than power

- After the ball is hit

  - FREEZE your stroke so that the racquet head stays up and the racquet face is still pointing toward your target. **This one action will improve your volleys dramatically,** as it forces you to keep the face of the racquet facing the target and prevents you from dropping the racquet – a nearly universal bad habit.

  - Follow the path of the ball toward the net to (1) make your next shot the best it can be and as close to the net as can be, (2) take time away from your opponent on your next shot, (3) intimidate and distract your opponent and (4) catch the ball before it drops below the height of the net.

○ ALWAYS expect the ball to come back and ALWAYS be ready for it.

IMPORTANT NOTE: statistics show that ¾ of all volleys are backhand volleys; make sure you practice those more than your forehand volleys – especially backhand volleys on shots that are hit directly at you.

TARGETS

Your target will depend upon where you are on the court – at the service line, at the net or poaching.

TARGETS FROM THE SERVICE LINE

Most volleys from the area of the service line, with the exception of high floaters, should be viewed as approach shots, setting you or your partner up for a put-away volley on the next shot. Trying to hit winners too soon in a point (particularly from the service line) is the main cause of volleying errors. The possible exception to this rule is that high floaters, if wide, can be hit down the line or at the net person. If down the middle, volley deep down the middle. If you own a low wide-angle volley from the service line, use it. Otherwise, practice it.

TARGETS FROM THE NET

There are several targets you can choose from:

○ The standard target: at the opposing players' feet, if they are at or inside the service line. BUT, how many times have you done this and had

the opponent reflex back a winner? Too many for me. One author has intelligently suggested aiming at the opponents' "toe line" – the imaginary line that runs from toe-to-toe of both feet and extends to both sides of both feet. The idea is to aim at the toe line, but a foot or two away from the foot. This is a great idea because the opponent has to reach down AND over to the side to retrieve the ball. But, by reaching, they can still return the ball with a half-volley. I suggest aiming at a point that is 2-3 feet to the left or right of the feet but aimed a foot behind the feet – which is a virtually impossible shot to return in the air and impossible to hit as a half-volley. You are really aiming at your opponents' "ankle line."

- Directly at the net opponent's body (or backhand) – particularly if they have a weak volley game

- Either of the side "T's"

- If both opponents are at or inside the service line, then the center "T" is a great target.

- If one or both of the opponents are at the baseline, then a soft angled shot is very effective.

- IMPORTANT NOTE: if both you and at least one of your opponents are at the net and the ball is above the level of the net, aim at the opposing net person. If they get it back to you, make sure you hit the ball right back at the same person as they are the closest to you, they have the least time to react and they are still recovering from their last shot.

TARGETS WHEN POACHING

When you are the net but are poaching or middle-jumping, then your targets are slightly different. The general rule is that, when poaching, you should volley in the direction you are moving. So that if you are poaching from the deuce court and going to your right, then the target should be the right side "T." This is not only the easiest target, but has the added benefit of adding the momentum of your body to the force of your volley. Further, to try to volley behind you (behind you to your left) would leave the entire half of the court you just vacated wide open for your opponent to hit a winner into.

LOB VOLLEYS

If you don't own this shot, learn it. Lobbing your volley over the head of your opponent/net person – particularly when both opponents are at net – is one of the most effective shots in the game. Because people do not expect it, it always has the element of surprise, it usually cleanly wins the point and is very deflating to your opponents, who just stand there and watch the ball go over their heads.

TOUCH VOLLEYS

Soft volleys are very effective when one or both opponents are at the baseline. A short soft volley is very difficult to reach and, if reached, results in a weak shot hit up to you or your partner at net. There are two such types of volleys: the "stop volley" and the "drop volley." Learn at least one of them. The "drop volley" is just

a soft touch shot. The "stop volley" is accomplished by loosening your grip so that the racquet head actually goes back when the ball hits it, absorbing most of the pace and causing a very weak shot landing just over the net.

The longest distance on the court you can make your opponent run is from the baseline to the net. Make them run that distance - often.

TIP:
Gigi Fernandez teaches to volley low, waist-high and high balls with the same follow-through at waist height. This gives your volleys more depth and more consistency.

STOMACH VOLLEYS

One of the seemingly harder shots in tennis is trying to volley the shot hit right at your stomach or your right hip. Handling both of those is quite easy. Handle the one at your stomach by using your backhand in front of your body. Handle the one at your right hip exactly the same way, but by also taking step to your right so it becomes a stomach volley.

THE "SERVE AND VOLLEY"

THE SERVE

In addition to the information provided above in the "The Server" section, keep in mind that, when employing the "serve and volley" pattern, try to keep the serve down the center "T" or at the returner's body to avoid the angled cross-court return. Most pros recommend that 75%-85% of all serves should go there unless a different location better exploits a returner's specific weakness. Also keep in mind that, often a slower serve with more spin is more effective than a harder serve because the slower the serve, the more time you have to get closer to the service line for your first volley.

THE FIRST VOLLEY

The first volley is the most important shot in the serve and volley sequence. YOU MUST DECIDE ON THE INTENDED LOCATION OF YOUR FIRST VOLLEY BEFORE YOU SERVE. If you do not have great control over your service location, then decide on two different volley targets dependent upon where the serve lands. Assume that most points average 6 shots – 3 by you and 3 by your opponent. Do not try to end the point with your first volley unless it is a high soft sitter which you can hit down the alley or directly at the opposing net person. Rather, your first volley should be no more than an approach shot and should be treated as such (see, "Approach Shots" section). It is only intended to set up your (or your partner's) next shot. So do not overhit it, no matter how tempting (unless you have a great swing volley). If you have hit a good "approach" volley, then you have a much better chance of ending the point with the next shot, having accomplished your goal of ending the point with 3 shots.

Do not run like crazy after your serve to get as close to the net as possible; run at a comfortable speed, maintaining total control of your body so you can split-step as your opponent hits the ball. If you volley from a foot further from the net it will have no impact on the quality of the approach volley. It is much more important to maintain your balance.

As an approach shot, keep your first volley either very deep or very short but *never* at or near the service line unless it makes your opponent run to get it. See Martina's TACTIC below. If in your arsenal, use a short-angled shot. You can hit it directly at the baseline opponent or angled away from them – but never toward their stronger side, without having to make them run. The worst shot you can hit is to their strong side with the ball bouncing at or

near the service line. From there, your opponent has a variety of aggressive shots they can make to take control of the point.

TACTIC: It is not a natural shot, but one worth learning: the serve and volley first volley as a lob volley over the head of the opposing net person. Very effective.

TACTIC: Plan to serve wide, drawing your opponent off the court and then hit your first volley down the middle or hit a lob volley over the net person. Conversely, serve down the middle, and plan your first volley to go wide.

MARTINA'S TACTIC: Martina Navratilova (who won 109 consecutive matches over a two year period with Pam Shriver and has the record for doubles masters' wins at 19 – three more than the Bryan Brothers!) strongly believes her best first volley was one that did not have much pace on it and was aimed at the center "T," with some slice so it stayed short and low. This shot forces your opponent to move forward and reach for the short low shot, making him hit the ball up to be put away by your team.

TACTIC: Anytime your net partner sees your opponent reach for a ball below the level of the net, they should immediately poach toward the center strap. Assume that the ball will be hit up and over the middle of the net.

PRACTICE: Everybody practices their volley from the net position, close to the net. Practice volleys from the service line or just behind it, as that is where most of your volleys will be hit and all of your first volleys off the serve and volley. See, PRACTICE section.

Also – when practicing with a partner, it is common to practice volleys cooperatively so that you are always aiming your volleys at your partner so that your practice partner can volley the ball back to you and you both get to practice your volleys at the same time. This is good practice if you want to practice always hitting your volley to the opposing net person. Instead, try taking turns being the "volleyer" and practice aiming your volleys toward your serve and volley targets.

THE HALF-VOLLEY

The best ever and clearest explanation of the mechanics of the half-volley I have ever seen was at the Peter Burwash Tennis Camp in the California desert. The instructor stood in front of a wall with a tennis ball in hand. He bounced the ball on the ground at an angle close to the wall so it then bounced up off the wall where he caught it. What became immediately obvious were several things:

1. The wall did not have to swing in order to get the ball back at a height well above the 3-foot height of the net;

2. The wall did not have to begin with any backswing;

3. The wall took no forward swing; and,

4. The wall was able to direct the ball forward away from it and toward its "opponent" and at a good height while it was absolutely perpendicular to the ground.

That clear demonstration provides the basic simple rules for effectuating a good volley:

1. Bend your knees so that the edge of the racquet is on the ground. It can always be raised if the ball bounces somewhat higher than expected. With your racquet on the ground the ball should never get under your racquet!

2. Keep the racket by, or in front of your front foot;

3. Open the racquet face to the extent you want to hit the ball higher and deeper;

4. Do not use your wrist to help the ball; keep it solid; and,

5. Upon impact, do not swing – either make no forward movement with the racquet (if you want to hit a short shot) or push the ball toward your target, leaving the racquet face at all times pointing toward the desired target.

GENERAL TACTIC

To hit the ball back to an opponent who is at the baseline, half-volley it either high and deep to the baseline or short, angled and low as a drop shot. NEVER aim a half-volley at the net person, as it will be an easy put-away for them (unless the opposing net player is sophisticated and is poaching when they see your racquet drop down for your shot).

SERVE AND VOLLEY TACTIC

When serving and volleying, expect the returner to aim the return for your feet. This is the hardest return to defend against and the one most commonly employed against the serve and volley. It requires the server to half-volley up so that a well-prepared and astute returner's partner will immediately poach when seeing your racquet dip below the net in preparation for your shot. The absolutely best shot in that situation is a short cross-court half-volley drop shot aimed at the returner's alley. It is easier to learn than it sounds and is a devastating and fun shot to add to your quiver.

NOTE: after hitting this shot, be sure to follow the ball to net with your racquet up, moving your feet and *expecting* (as always) your opponent to get to the ball and return it.

THE OVERHEAD

ELEMENTS

1. FEET, FEET, FEET: Constantly move your feet to get to the ball.

2. IMMEDIATELY both partners need to call who is going to hit the overhead by shouting "mine" or "yours." The partner hearing this should acknowledge the choice by shouting "ok" or some similar verification of who is going to hit the ball. NOTE: this not only clarifies who will be taking the overhead, but by shouting "mine," many people feel more confident in hitting the overhead. An interesting psychological boost.

3. IMMEDIATELY turn sideways and put your racquet up to prepare; take side steps to shuffle back to the ball quickly. It is far safer and quicker than trying to back up to get to the ball. Then take tiny adjustment steps to put yourself in perfect position.

4. Move back behind the point of contact, so that you can then step forward into the ball as you are hitting it, putting the momentum of your moving body behind the shot.

5. Point to the ball with your left arm straight and your left index finger aiming at the ball. Hold this as long as you can.

6. Likewise, as with your serve, keep your head up and your chin up as long as you can.

7. Reach for the ball to try to hit it at the highest contact point.

TACTICS

You can't play good doubles without being at net, and you can't be at net without a good overhead. Your objective should be to *never* miss an overhead. This isn't as impossible as it sounds if you employ the following techniques:

- PRACTICE: few people practice their overheads. You should practice them as much as your serve.

- SHORT LOBS: those you can attack – but by attacking, you don't have to crush them - no matter how good it makes you feel. The advantage of getting a short lob is that it forces you to get closer to the net. From there you can easily angle the overhead into the alleys so it bounces well off the court. Make it a "touch" shot - not a blistering example of how hard you can hit the ball.

- MEDIUM LOBS: those that are deeper than the short lobs, but ones that you can still easily get to around the service line. Again, do not try to blister these. Rather, hit a well-controlled overhead down the center of your opponents' court between the baseline and the center "T." This gives you the biggest target on the court and splits your opponents. Wait for an easier shot to put the ball away.

- DEEP LOBS: if you have to run back for them, sidestep to get to them. If they are deep or you are in anyway off-balance then do not attempt a normal overhead. Rather, pop the ball high and deep back to your opponents. This is like a high volley, but the angle of the racquet face will be pointing more up toward the sky. Practice this shot. It is not a natural stroke, but can be learned easily and quickly. It is a defensive shot intended only to get the ball back with enough height and depth to challenge your opponents and allow you and your partner time to recover your positions.

- PROFICIENCY IN DEEP AND MEDIUM LOBS: If you have a very good overhead and can aim it well, then feel free to take a full swing at the lob from the middle or back of the court. However, remember that, because you are farther from the net, you need to change your stroke slightly by swinging more out than down. To accomplish this, aim for the base of the fence rather than your usual target in the court.

- BACKHAND LOBS: if you are very close to the net, you can softly drop them over the net. If they are any further back, follow the same procedure as you would for a deep lob, keeping the racquet face pointing up the entire time and aiming high and deep cross-court, where you have more court to hit into.

- PRACTICE; PRACTICE; PRACTICE

FOLLOW-UP TACTIC

If your partner has a strong overhead, plan ahead with him so that, immediately after your partner makes contact with his overhead, you move to the net to mirror the ball and put away any weak reply, with your partner remaining deeper to protect against another lob.

WHICH PARTNER TAKES THE LOB?
Easy: The one who has the easiest shot to hit with a forehand overhead – or, if there is time to choose, then the one with the significantly better overhead.

DEFENDING AGAINST THE OVERHEAD

Most people are afraid of getting hit by the ball – especially an overhead at close range. If your reaction is to turn around, you will never return the ball! There is a very specific method for dealing with both the overhead and any high close-range "power" shot hit by your opponents:

As soon as you see a situation develop where your opponent has a short overhead, expect the ball to come at your feet. Immediately, pedal backwards as quickly as you can. As soon as your opponent begins to swing, squat down as low as you can with your feet wide apart (like a hockey goalie) and your racquet as low as possible and in front of you. Do not swing at the ball when it comes at you. There is no need to as it will have enough pace so that if you just firmly put your racquet in the path of the ball, it will go back over the net. Practice this shot. You will be AMAZED at how many of these shots (even if aimed right at you) that you will be able to reflex back.

APPROACH SHOTS

RULE #1 IN TENNIS: GET TO THE NET! Unless you are already playing the net position or are running up toward the net to retrieve a short ball, then the only way to get to the net is by hitting a transition shot – an approach shot – as you approach the net. There are several flavors of approach shots:

- The standard approach shot

- The chip-and-charge

- The first volley off of a serve-and-volley

- A "sneak approach"

THE STANDARD APPROACH SHOT

This shot is utilized when you have received from your opponent a relatively short shot that lands within four feet of your service line. This shot is an invitation for you to move forward to hit the ball so that, in doing so (1) you approach the net and (2) you take their weak shot and hit an offensive shot that forces your opponents to return a weak shot that can then be put away by either you or your partner, both of whom are at net.

EXECUTING THE SHOT

As you move toward the short ball, do not stop to hit it. Rather, move *through* the ball so that you are moving forward as you hit the ball. This is called 'karaoke" or cross-over footwork and it gets you closer to the net and adds your body's forward momentum to the pace of the ball. Your object is to hit a shot that forces your opponents to be defensive. Your standard choices include:

- ➋ Deep down the center "T" – particularly effective if the opposing net person never poaches;

- ➋ Short slice down the center "T;"

- ➋ Any deep slice or topspin that makes the opponent move backward or run sideways.

IMPORTANT! The approach shot is exactly that: a shot used to approach the net to then volley a winner on the next shot. **IT IS NOT INTENDED TO BE NOR SHOULD IT BE A WINNER.**

MORE CREATIVE APPROACH SHOTS INCLUDE:

- ➋ The lob;

- ➋ The drop-shot; no one ever expects this shot as they see you "approach" toward the net and so it is very effective. It also forces your opponent to hit up on the ball should he even reach it;

- ➋ A shot directly at the net person or at their weaker back-hand to cause a weak return.

THE CHIP-AND-CHARGE

A very effective shot off of a short ball (usually a weak second serve) that requires you to move into the ball, hitting it at waist height or higher (usually with slice) and aimed deep to the baseline or to the side "T." As you are close to the net and are closing quickly, it can also be aimed at the net person or down-the-line.

THE FIRST VOLLEY OFF OF A SERVE-AND-VOLLEY

This is discussed in detail in the SERVE AND VOLLEY section.

THE "SNEAK" APPROACH

This can be used at any time with any shot you hit that places your opponents on the defensive. It can either be planned or spontaneous. The goal is to hit a shot that is intended to place your opponents in a weak position. After they have chased the ball down, when they are about to hit it and are not looking at you – you sneak up to the service line or net. By surprising them, you will get a ball that was intended to be hit to you at the baseline and should be easy to pick off. The surprise factor alone is significant, as your opponent will not expect to see you at net and your presence there will take away their time in reacting to your next shot. If they do happen to notice you sneaking in, then they will have to quickly change their intended shot, many times causing a mishit.

THE POACH & THE FAKE POACH

We have dealt with the positions of the SERVER'S PARTNER and the RETURNER'S PARTNER sections of this book that cover starting positions and movement patterns available to those positions. Please remember that the net person should always be starting in different positions and should always be moving - preferably in different patterns. Here, we deal with the mechanics and timing of two of those main options: the poach and the fake poach.

THE POACH

Nothing is more fun in tennis than picking off one of your opponent's shots and putting it away. It is worth way more than the one point you get for hitting a winner. In addition to hitting a winner, you will forever during that match make your opponents wonder if you will or will not poach – and, by adding changing starting positions, changing patterns and fake poaches, you will actively interfere with the returner's concentration, make them miss shots down-the-line, focus on you instead of the ball and change their planned return just before they hit the ball, resulting in mistakes and mishits. These factors apply not just to the return of serve, but even more so to poaches that occur during points. The secret of the effective poach is timing. There are two different ways to time your poach:

THE SURPRISE POACH

The more common one: This poach is performed when you do not want your opponent to know that you will be poaching. Therefore, you must not begin your poaching movement until you know that your opponent is not looking at you and cannot see you – either (1) when your opponent looks down at the ball that they are about to hit or (2) when your opponent starts the forward swing of his racquet to hit the ball. Then, and only then, do you start your movement.

THE DISTRACTING POACH

The other method is to begin your poach just a second or two before the "surprise poach", so that your opponent sees you move just before they start their forward racquet movement, and makes a last second effort to avoid you by trying either to hit down the line or further cross-court out of your reach – causing them to instantaneously change their target and miss the new target. The usual time to begin this poach is just after the ball lands in your opponents' court.

THE PATH TO THE BALL

As discussed in the section "MOVE DIAGONALLY," you must not move laterally parallel to the net. Rather, you MUST move at a 45-degree angle toward the ball and the net strap at the same time – allowing you to cut off the ball earlier, get to more balls and provide bodily momentum to the volley you hit.

TARGETS

Again, this topic is covered in the section on VOLLEYING. It is important to add, however, that all pros agree that you should volley the ball in the direction you are moving because doing so (1) provides bodily momentum to the shot and (2) hitting back toward the direction you came from leaves that entire half of the court wide open for you opponents to hit the ball into.

COMMUNICATION WITH YOUR PARTNER

Poaching can be prearranged by agreement with your partner (either by hand signal or verbal agreement), preplanned by you and your partner based upon the location of balls hit by your team, preplanned by you without your partner's knowledge or completely spontaneous by going for a ball you had not planned to poach but your eyes light up because you know you can get to it and put it away. In all of these situations, it is important to know and for your partner to know, what she should do in the backcourt when and if you poach. If the poach is preplanned with your part-ner, then your poach will carry you past the center line and your partner will cross over to the side of the court you just vacated. If, however, your poach is one of the other types, then it is import-ant for you to only go to the center line to see if you can pick off the ball. Your partner will assume that you will return to your half of the court unless you believe that, having gone to the center line you can go further to get the ball, in which case your partner should know to switch sides.

THE UNPLANNED POACH

A misnomer really – because you should *always* plan to pick off a ball you can get to and put away, even if you haven't planned to move all the way to the center strap.

BE HUNGRY! ALWAYS WANT THE BALL TO COME NEAR YOU SO YOU CAN PUT IT AWAY.

PREARRANGED POACH

If you and your partner want to use prearranged poaching, then you must decide whether to communicate by signal or verbally. There is actually a big difference as, when signals are used, it is the net person directing where the serve will go and whether the net person will poach or not. On the other hand, during verbal communication, it is usually the server who makes that decision. The benefit of verbal communication allows the partners to discuss it. The verbal method also allows the partners to decide ahead of time the strategy for both the first and second serve.

THE FAKE POACH

The fake poach should be employed almost every time you do not poach. Your intention is to distract and confuse your opponents and/or to make them think that you are poaching.

IMPORTANT NOTE: <u>YOU CANNOT EVER SUCCESSFULLY FAKE A POACH UNLESS AND UNTIL YOU HAVE PROVEN TO YOUR OPPONENTS THAT YOU CAN AND WILL POACH!</u> Otherwise, all the best movement and faking in the world is totally useless. Why should I ever think that you are going to poach if you never do so?

The purpose of the Fake Poach is exactly the opposite of the poach and therefore the timing of its execution is also opposite. Instead of hiding your actions so as to surprise your opponent, you want them to see you move and to think that you are poaching – or, at the least, question whether you are poaching. The timing must therefore be different. Because you *want* to be noticed, begin your movement of "poaching" with the same movement as if you were actually poaching, but start it earlier – just after the ball

bounces in front of your opponent and *before* they look down and begin their swing. As they notice you start to move, they hopefully will change their mind about their target and, as well, be distracted by the movement. The fake poach can and should be employed throughout the point – not just as the server's partner on the return of serve. ALWAYS move in an unpredictable manner just as your opponent is about to hit the ball.

THE MOVE: The fake should be noticeable but not overly dramatic. You don't need to hop and jump and move all around. Rather, take one step toward the center line and make one large stretch with your racquet toward the center line. Then recover to your original position. Some players add a shuffle with their feet so as to create noise distraction as well as visual distraction.

TACTIC: Should your opponent have a short ball that can easily be hit down-the-line, instead of covering the line, move a step in the *opposite* direction toward the center line and fake moving even further. As the opponent is just about to hit the ball, quickly jump to cover the line. This move can be enormously successful if you time it correctly.

TACTIC: As your desire is to be seen, it is good to make even more dramatic of a first move with your foot and your racquet than if you were actually poaching.

PRACTICING

Ask yourself this question every day:

"WHAT CAN I DO TODAY TO GET BETTER?"

Take heed of Alex Rodriguez's advice: "You never stay the same. You either go one way or the other."

Practice weekly and practice seriously. Most club teams have a "practice" that consists of the team dividing up and playing each other. That is NOT practice! Use your practice time to make sure you practice:

- Second serves: there is the old saying that "a player is only as good as their second serve." There is some truth to that. A weak second serve gives the returner a psychological advantage and allows them to attack your serve. A stronger second serve not only negates that, but gives you the option of feeling freer to try stronger first serves, knowing that you have a good second serve to rely on.

- Returns of serve: the first or second most important shot in tennis. More points are lost on return of serve than any other shot. PRACTICE THEM! Very easy to do – just find someone who wants to practice their serve; and, if possible, find a 3rd person to act as a hungry server's net partner.

- Lobs;

- Volleying the ball out of the air;

- Volleys from the service line; most volleys you make, and all first volleys in your serve-and-volley sequence are from or slightly behind the service

line; and yet, very rarely do people practice those volleys;

- Volleys in general: practice "freezing" the racquet after every volley to insure that you develop the habit of not dropping it after or during the shot. To help practice this, try practicing all volleys by holding onto the racquet with your left hand on the throat of the racquet and not letting go. This will force you to turn for the shot, keep the racquet head up before, during and after the shot, not take a backswing, not take a big forward swing, and keep the head of the racquet facing the intended target.

- Backhand volleys: statistics show that 75% of all volleys are backhand volleys; practice them more than forehand volleys – particularly volleys that are directed right at your body (see below).

- Volley targets: most volley practices are cooperative "volley-to-volley" drills - however, by doing so, you are practicing hitting your volleys to your opponent rather than at winning targets. Refer to the "VOLLEYS" section in this book, learn what the best targets are and practice hitting to them so your reflexive action is to always hit toward a target.

- Soft volleys;

- Drop shots; and

- Overheads.

PRACTICE TIP

Rohan Bopanna, the former #3 doubles player in the world, runs a tennis camp in India. One of his best ideas is to allow his players only one ball during hitting sessions. It is astounding how one's natural disinclination to stop to pick up a ball can serve as a motivation to never miss a shot. A great idea for practicing consistency and concentration. Try it next time you rally or even warm up.

THE "RDC" VOLLEY DRILL

Bob and Mike Bryan learned this drill from the Romanian Davis Cup ("RDC") team when they watched that team practice years ago. Both players start on the service line (or slightly inside) on opposite sides of the net in opposite alleys and volley back-and-forth to each other. After each hit, both players move a step or two toward the opposite alley, making sure to feed the ball to their partner just ahead of them – that is, where they will be after their next step rather than where they currently are. Both partners should reach the opposite alleys at the same time and then reverse toward the other ally, all the while continuing to volley back and forth.

See how many times back and forth you can go. A GREAT drill for volley footwork, concentration and hitting volleys from different directions!

THE "KAMIKAZE" DRILL

Stand 4 feet from the net. Have someone stand across the net on the service line and fire balls at you hard and quickly for you to volley. Or, use a ball machine on a relatively "high" speed. It will force you to "tighten up" your volley by eliminating unnecessary movement in your stroke and will sharpen your reflexes. You will get used to this. Then turn the speed up to high. If you get used to this, then move closer to the net and, if necessary, you can move the ball machine closer to you.

This drill was used successfully by Tim Gallwey and the interesting results are spelled out in detail in his book, The Inner Game of Tennis," Gallwey, Random House, 2008 at pages 94-95. Students adjusted to the high speed, and then re-adjusted to the even higher speeds and not only found their reflexes improving, but also their form and their being in the "present moment."

SERVES

Lleyton Hewitt once demonstrated the way he practices his serve: use 4 cones to divide both service boxes into thirds – representing wide, center line and middle (body) serves. Practice *second serves only* and do it by making yourself serve 5 balls in a row wide into the deuce court box, then 5 to the middle of the box and then 5 down the center "T." Do not move to the next target area until you have completed 5 in a row. After completing 5 in a row to all 3 deuce court targets, then move to serve to the AD court, repeating the same process. Do this repeatedly, and you will see an improvement in both your first and second serves. READ: "Muscle Memory and Imagery: Better Tennis," by Archie Dan Smith. Very effective.

For professional help with your service motion, check out Dr. Mark Kovacs, who has a PhD in biomechanics and specializes in tennis instruction – having coached many big name players. See, kovacsacademy.com.

DROP SHOTS

Practice drop shots by putting a garbage can on a chair just on the other side of the net on the singles sideline so that the top of the can is just below the top of the net and the can is not more than 6 inches from the net. Most mistakes in hitting a drop shot occur because (1) there is not enough of an arc on the shot and (2) because the apex of that arc needs to be on *your* side of the net – so that, as the ball crosses the net, it is descending. It is impossible to get the ball into the can unless you get a good arc that reaches its apex on your side of the net.

Then try it without the chair and can. A successful drop shot is defined as one that bounces close to the net and then bounces at *least* twice before reaching the service line. A great drop shot is one that never reaches the service line. But don't sacrifice getting the ball over the net for perfection. The drop shot should not be intended as a winner – just another approach shot in your arsenal that causes your opponent to run and reach. And don't forget to charge toward the ball after you hit it so that (1) you are prepared for the ball to come back, (2) you cut off the angles your opponent has to return the shot and (3) you intimidate your opponent.

In order to get the ball to stop moving forward once it bounces, you need to add backspin to the drop-shot. If you are having a hard time with this, have someone throw a ball to you. Catch it gently with your racquet so that the ball spins with backspin straight up in the air and then catch it with your left hand. Do this and you will quickly develop a feel for the drop-shot backspin.

SMOOTH YOUR STROKES

Do you want to practice effortless ground strokes and volleys? Try practicing with your partner using low compression tennis balls. They come in different degrees of compression. Buy at least two types. Try them both to see which one you like the best. Hitting with them removes the feeling that you are "trying hard" to hit a shot and gives you a gentle, natural and effortless stroke. Do it enough and you will incorporate it into your game.

PRACTICE GAMES

1. Play with 4 people and play a standard game but start with the score at 30-30 or deuce. This will force all 4 players to focus intently on the first points that are played, practicing a high level of concentration under pressure.

2. Play any game or practice drill, but, in addition to giving a point to the winner of the point, subtract one or two points from the loser if the point is lost as a result of hitting the ball into the net.

3. Play just 7 and 10 point tie breakers. More pressure and it gives you practice playing tie breakers.

IMPROVEMENT AND PLATEAUS

Studies that chart the way people learn and the process they go through demonstrate that the learning "curve" is not a curve at all. Rather, it takes a while for any learning to be adopted into one's playing; and then, the quality of the playing dramatically increases to a high point, then drops down a bit and then stabilizes on a plateau for a long while. So this is why we get frustrated when we do not improve quickly and dramatically after practicing a lot or trying to learn a new shot. It also explains why we get better and then get a little worse after reaching the apex. By understanding our learning "curve" we no longer need to be frustrated. Just learn, practice and practice more – without frustration when we stay on the same plateau and do not see improvement – because it will come in a big way. Patience.

VISUALIZATION

Mental and psychological experts who study learning (including sports learning) have determined that you can learn and practice many things without actually physically doing them – but by mentally performing the actions in your head as if you were physically doing them. You can, in fact, improve not only your strokes by imagining a perfect stroke, but you can also perfect your tactics by imagining different scenarios in your mind and practice your response to them – over and over. Dr. Mark Kovaks, a former tour level player who holds a doctorate in sports biomechanical engineering and who coaches many high-level players has even suggested playing an entire set in your mind! Grigor Dimitrov, a top mens' singles player, was asked how he practiced during the Corona Virus lockdown. He responded by saying that he randomly selects an opponent from the list of top players, studies that player and plays an *entire match* in his head. He says emphatically, "It works!" Playing points and games in your head will improve your concentration, focus and shot selection.

PRACTICING WHILE INJURED

For the past 5 years, I have been unable to play matches 50% of the time due to injuries. But as soon as I am able to stand, even without moving much, I get the ball machine, stand in one place and practice drop-shots. As soon as I am able to take a few steps, I practice slice forehand returns of serve. You can stand at the net with your practice partner on the other side and just practice mini-shots just over the net. There is always something you can do to practice – if not on the court, then with visualization, reading, on-line courses, videos, etc.

BEST TIP

Find a great teaching pro – not a good one, but a great one – preferably one who specializes in doubles. Take doubles-oriented lessons from as many other good pros as you can find, as every pro
has a different eye and will spot different needs in your game.
Different pros have different ways of explaining the same things
– some of which are better suited for some people and others for
other people. As well, different pros will be able to share different tactics. Spend your time working on shots that are particularly
beneficial in doubles, and practice doubles tactics and strategies
more than you practice strokes.

CONCLUSION

- LEARN
- PRACTICE
- PROBE FOR WEAKNESSES
- ATTACK WEAKNESSES
- BE POSITIVE
- LOB
- APPROACH
- SERVE AND VOLLEY
- GET TO THE NET
- STAY AT THE NET
- COVER THE MIDDLE
- CONSTANTLY MOVE AT NET
- POACH
- FAKE POACH
- DON'T OVERHIT
- KNOW YOUR TARGETS
- HIT YOUR TARGETS
- INTIMIDATE
- WIN
- HAVE FUN

I hope you have as much fun learning and using these tips and tactics as I have had. Enjoy the process of learning and executing what you learn.

~ Bob Allcorn (DOUBLES DOMINATION.NET; allcorn@comcast.net)

GAME-TIME Rx

Having issues during a match? Missing shots? Not playing your best? Make a copy or take a photo of this page and keep it with you to read during change-overs. (there is a second identical page after this one that you can remove to put in your tennis bag).

GENERAL TIPS

- FOCUS ON THE BALL; focus on a small part of the ball or on the color on the ball;

- Concentrate only on moving your feet;

- Use "bounce-hit" method explained in ATTITUDE section

SERVE PROBLEMS

- KEEP THE TOSS HIGH

- Keep chin up and left arm up as long as possible

VOLLEY PROBLEMS

- KEEP RACQUET UP AND AS FAR IN FRONT AS YOU CAN KEEP IT

- Move feet toward ball

- "FREEZE" your racquet after the shot to insure that the racquet face points toward target

- Aim for targets vs. using power

GROUND STROKES

- KEEP KNEES BENT

- KEEP HEAD DOWN UNTIL SHOT COMPLETED

- Step into the ball

- Focus on lots of net clearance.

OVERHEAD

- MOVE FEET – A LOT

- Stand side-ways

- Keep chin up and left index finger pointing at ball until contact

- Aim for just behind the center "T"

ATTITUDE

- Slow down between points; take time; re-tie your shoe laces

- BREATHE (see ATTITUDE section)

- Act confident and "up"

- smile

THE AUTHOR

I was only at Center Court at Wimbledon once – sitting in the 5th row! I am not a tennis pro and never have been. I spent 40 years as a recognized expert in commercial litigation practice. 15 years ago, I was asked by the country's largest legal publishing company WEST PUBLISHING CO. (now Thomson Reuters Publishing Co.) to author a legal manual on Litigation Tactics. After two years of work, I completed a treatise of 800+ pages.

As a passionate devotee of excellence and gaining tactical advantages against adversaries, I have always been keenly aware of, sensitive to, studied and searched for ways to dominate my adversaries – through exercising tactical advantages, knowledge of the details of the Court Rules, identifying weaknesses in my opponents and knowing how best to attack and exploit those weaknesses. At the same time, I have also been a student of tennis – especially doubles. Although the aging process, a torn hamstring, a knee replacement, two hip replacements, two broken ribs, one bout with cancer and the usual battles with tennis elbow, plantar fasciitis, etc. have resulted in my demotion from a 4.5 to a 3.5 player, I have always strived to educate myself as to the finer points of the "game" of doubles - and it is a game, the rules and elements of which need to be studied and analyzed.

To create this book, I have brought to bear on the subject of doubles tactics the same laser focus I employed as a litigation tactician. I have reviewed all of the notes I have taken throughout many years of lessons, clinics, camps, just about every book that has ever been written, most all of the videos that have been produced (including DVD's and even VHS tapes) as well as contemporary webcasts, websites and on-line tennis sites.

As my wife and I love to play mixed doubles as a team, we have travelled the world taking lessons especially devoted to doubles

strategy. I have picked out the best tips and tactics and included some of my own. I have tried to incorporate them all into my doubles game and have tested them all. From these, I have selected the very best of the best TACTICS and tips to include here for your consumption and education. They work. Try them.

Please enjoy learning them and using them as much as I have.

ACKNOWLEDGMENTS

There are many tennis pros around the world that I have been fortunate enough to have had as teachers and gurus. To all of them I give my eternal gratitude – not just for the education but for the pure pleasure and thrill of learning and playing this wonderful game. I cannot begin to mention them all. As always, though, there are those special few that, because of their tennis intelligence, experience, expertise and dedication to both the art of tennis and the art of teaching, deserve special attention. And to those I give my wholehearted thanks and appreciation for all that they have done and continue to do for me and the rest of the tennis world:

VIMAL PATEL: former #2 U.S. 18 and under; 1988 Kentucky High School State Singles Champion; 1989 Florida High School State Singles Champion; 3 time All-American at Stanford; 2 time National Doubles Champion; former singles and doubles touring pro; head teaching pro at Tough Tennis, Naples, Florida; Inductee to the Kentucky Tennis Hall of Fame.

ANDY YENAWINE: Head teaching pro at Center Court Tennis Academy

PETER BURWASH: former professional hockey player; former professional tennis player; Olympic Gold Medal Winner in doubles; USPTA Master Professional tennis Instructor; member, USPTA Hall of Fame; tennis announcer; founder and CEO of Peter Burwash International, Inc, the manager of 112 professional instructors at 59 locations in 24 countries as well as PBI World Tennis Camps – and all of his professional instructors that I have had the pleasure of learning from.